HEAT TRANSFER

Dimitri Gidaspow

Heat Transfer

INTRODUCTION

There is a need to construct buildings with small loss of energy. Such buildings were constructed in China by Professor Leon R. Glicksman of MIT (L.R.Glickman and J.H. Lienhard, Modeling and Approximation in Heat Transfer)

We define work as W ,Pressure as P and displaced volume as dV. Then we obtain dW = PdV.

HEAT EXCHANGERS ENERGY BALANCE

An energy balance between a hot fluid and cold fluid is as follows:

Net rate of heat outflow by convection + rate of heat transfer between 2 streams = 0.

$$C p \ d T/dx = a \{Tc - TH \}$$

where

C p is the specific heat of the fluid , T is temperature , x is the distance in the direction of flow and a is a constant.

Integration of the above differential equation with respect to x gives the logarithmic expression shown below for the case Cph not equal to Cpc

$$\ln \{[\ Tc - TH \ ]/[Tc(o) - Tc(0) \ ]\} = \ ax \ /Cp \ (I - Cph/Cpc \ )$$

The above expression can be written in terms of the logmean temperature difference given on page 463 in Transport Phenomena 2002 book by R.B.Bird, W.E.Stewart and E.N.Lightfoot.

Heat Transfer

SOLAR ASSISTED AIR CONDITIONING OF BUILDINGS

Hans-Martin Henning in an article Applied Thermal Engineering 27 (2007) 1743-1749 explained how to use solar assisted air conditioning to improve the performance of commercial and residential buildings. The earliest related paper on the subject is the Munters environmental control system by W.F.Rush and R.E.Macris.Their system consists of 2 rotating heat exchange wheels. Later at IIT in Chicago we build a cross-cooled desiccant dehumidifier (W.W. Worek and Z.Lavan, J.Solar Energy Engineering, Aug.1982,Vol. 104/187).

In a presentation at the 13th ICEC Conference in August 1978 Gidaspow showed the structure of the silica gel sheets we used in our experiment using an electron microscope. The earliest work on the subject is that of N.E.Pennington, Humidity Changer for Air Conditioning, U.S. Patent 2,700,537 ,Jan.25,1955.

In our paper with Worek and Lavan, we used the following equations to modeling:

Adsorption

Process Stream: $\partial Y/\partial x = Kc/h(Yw - Y)$

Wall: $\partial w/\partial t = Kc/h(Y-Yw)$

Energy Balance

Process Stream: $\partial T/\partial x = Tw - T1$

Cooling Stream: $\partial T/\partial y = Tw - T2$

Wall: $\partial T/\partial t : (T1 - Tw) + b(T2 - Tw) + QKy/h(Y- Yw)$

Desorption

Mass Balance

Process Stream: $\partial Y/\partial x = Kyc/h1(Y - Yw)$

Wall: $\partial w/\partial t = Kycw/fh1 (Y - Yw)$

Heat Transfer

Energy Balance

Process Stream: $\partial T1/\partial x = Tw - T1$

Wall: $\partial Tw/\partial t = (T1 - Tw) + QKy/h1(Y - Yw)$

Purging/Preheating

Energy Balance

$\partial T2/\partial yp = Tw - T2$

Wall: $\partial Tw/\partial t = (T1 - Tw) + QKy/h1(Y - Yw)$___

ENTROPY PRODUCTION

The rate of entropy production is a method to obtain the best possible design.

Let S be entropy, P the pressure and L the length and m the mass rate of flow.

Make a balance on a pipe of length L. For an adiabatic pipe of length L and a

steady flow of mass rate m, the entropy production due to friction = m(Sout − Sin)

From thermodynamics we know that dH= V dP + TdS from the definition of enthalpy per unit

Mass H. For adiabat flow dH = 0. Then TdS = - VdP .

We can express the frictional loss by means of a Fanning equation, F = 2fVVL/D where D is the pipe diameter and L is the pipe length and f is the friction coefficient. Then the entropy production per unit length = m/T (2fVV/D).

The entropy production can also be expressed in terms of the Reynolds number =

Heat Transfer

4m/PD viscosity.

We find that the dissipation rate increases as the cube of the flow rate and inversely as the square as the fifth power of the pipe diameter and inversely as the square of the dDensity for a given flow rate. It is proportional to the friction factor.

Entropy Production due to Heat Transfer

Consider a pipe length of length L but now allow heat transfer to take place through a film coefficient J/mm-sec-deg K.A hot fluid flows a space of Temperature T and T + T +delta T.

For a steady state flow the rate of entropy production = net rate of entropy outflow due to heat transfer = Q/T – Q/(T+delta T).

For delt T much smaller than T, the rate of entropy production = delts T Q/TT.

In terms of Delta
T,

The entropy production = film coefficient X PL square delta T/square T.

Entropy Production due to Heat Transfer, Mass Transfer and Friction

Since the entropies for heat transfer and friction are additive, we obtain the sum of the two   equations that show that there exists a maximum value. The rate of entropy production decreases with diameter due to friction but increase due to heat transfer. Therefor there is an optimum point. The rate of mass transfer can be added as an exercise.

Heat Transfer

Carbon Dioxide Capture of CO2 from Power Plants

Dr. Hossein Ghezel-ayagh, my former Ph.D. Student at IIT, has developed a method of removing CO2 from power plants and converting the CO2 into nitrogen and oxygen. He has done this at FuelCellEnergy a company founded by Bernard Baker who obtained his PhD at IIT under my direction. The CO2 is captured and converted into nitrogen and oxygen in a molten carbonate

fuel cell developed at IIT under supervision of B.Baker. The description of the fuel cell is given Chemical & Engineering News, March 29,2021, p 18-19.

Flue gas (10%CO2 + O2,N2) enters the cathode of the fuel cell at the bottom and leaves the cathode as nitrogen and oxygen. On top methane and water vapor enter the fuel cell and leave as 70% +H2O,H2. It is impossible to use this method to remove CO2 from air.

Carbon Dioxide Capture from Air

We are developing a method to capture carbon dioxide from air based on our patent U.S. 3,865,924; Feb 19, 1975, "Process for Regenerative Sorption of CO2 " D.Gidaspow and M.Onischak. We used this method to remove CO2 from nuclear submarines.

We are building a small unit to obtain data in Thailand under the direction of Professor Benjapon who is working in their top university.

EXERCISES

1. Transpiration cooling in a cylinder

Heat Transfer

Consider flow in an annular cylindrical porous insulation of length L and radius R1 and R2 maintained at temperatures T1 and T2 ,respectively. Dry air is blown outward radially at a velocity v m/s. The density of air is d kg/s.

   A. Formulate the steady-state conservation of mass equation for the gas in the annular   space. Find w, the mass rate of flow of the gas, the density, the porosity and the cylinder radius r.

   B. Formulate the steady state conservation of energy equation for conduction and and convection in the annular space in terms of temperature T.

C. Let R0 = wC/2 py kL. Show that the solution to the differential equation in part B is as follows: T = c/Ro + C2/Ro where the c and C2 are integration constants.

   D. Find the constants of integration.

   E. Find the heat flux at the inner radius

F.   Consider the limit of gas flow, that is conduction in the cylinder held at the specified temperatures. Find the heat flux at the inner radius in terms of the given temperatures and the cylinder radii. Show how this can be done in two different ways.

2. Combustion in a Cylindrical Porous Burner

Air and fuel are fed at a rate w kg/s radially into a cylindrical porous burner.

The rate of heat generation due to combustion of the fuel is R kw/cubic meter.

The surface temperature is constant.

A. Formulate the differential equation for the temperature distribution in the burner in terms of w and the effective thermal conductivity k.

Heat Transfer

B. Consider the limit of zero k. Find the temperature distribution in the burner for a constant rate of heat generation.

C. Consider the other limit, that of zero w. Find the temperature distribution for a constant rate of heat generation.

3. Maximum Work from 2 Bodies

What is the maximum amount of work that can be extracted from 2 bodies initially at temperatures T1 and T2 ?

Their equations of state are: $U = NC_vT$ where U = internal energy,

C_v = molar heat capacity, N = number of moles, T = temperature.

HINT: Make an energy balance in terms of the final temperature Tf .

Find T.

4. Entropy Production due to Mixing

Component A flowing at a rate of ma moles/sec is mixed with component B which flows at a rate of mb moles/sec. Find :

(a) The entropy production due to mixing, neglecting heat of evolution.
(b) The entropy production with a pressure drop Δ P Pascals, assuming the mixture is incompressible.

(c) The entropy production with a heat leak of q kg/cubic meter.

(d) What is the relation of the entropy production in part (a) to the minimum work needed to separate the mixture into components A and B.

5. Availability Loss in Cooling Water

Heat Transfer

Cooling water leaves the condenser of a power plant at a temperature Tc and is thrown into a large lake at a temperature To. Fresh cooling water is taken from the lake at the same flow rate m kg/sec. The specific heat of water is a constant equal to Cp J/kg-deg K. Find the loss of availability of hot water dumped into the lake in terms of m ,Cp ,To and Tc.

6. Thermodynamic Efficiency of Distillation

A distillation column separates a 50-50 mixture of propane and propylene into two products , each 99% pure. The column is heated with steam at 212 deg F and cooled with water at 70 deg F. The entropies of the feed, the distillate, and the and the bottoms have been calculated using the equation for the mixture :

$S = -R$ Sum of (xi ln xi).

Steam enters the column at 24,600 Btu at 70 F at the bottom. The feed enters the center of the column at a rate of 1,377 Btu. The distillate leaves the column on top 99% pure at 70 F. The cooling water on top of the column is 70 F.

(a) Calculate the entropy changes for the steam and the cooling water, Assuming negligible temperature rise.

(b) Make an entropy balance and calculate the thermodynamic efficiency.

How can you make this column more efficient?

7. Second Law Analysis of Coutercurrent-Cocurrent Membrane Mass Exchanger

Heat Transfer

Consider hydrogen separation through a membrane and transfer to a carbon dioxide stream. Assume that hydrogen obeys the ideal solution theory and that the temperature is To.

(a) Find the minimum work necessary to separate hydrogen from the inlet mole fraction of x H22.

(b) Find the entropy production and the availability consumption due to a pressure drop of delta P Pascals along the length L of the hydrogen-nitrogen stream.

(c) Find the entropy production due to a pressure drop delta P Pascals Along the length L of the hydrogen-nitrogen stream.

(d) Formulate a linear flux-force relation for the membrane flow and diffusion in terms of delta P and delta x hydrogen. Also find the net entropy production for the exchanger and explain why countercurrent operation is desirable.

8. Friction Factor Concept

Friction, F can be defined by means of the steady state mechanical energy balance

vdv + gdz +dP/rho = -dF -dW

where F acts as the energy that has added to the Bernoulli equation to make it exact. For the pipe flow F is expressed in terms of a friction factor defined by F = 2fvvz/D.

NOTATION	S entropy
D pipe diameter	T temperature

Heat Transfer

f friction factor   v velocity

F friction

g gravity

P pressure

Q heat into the system

W shaft work

X weight fraction

z height

u internal energy

(a) Use the steady state energy balance to show that

d(u+P/density + vv/2 + gz) = dQ-dW

(b) Consider flow of a multiphase mixture containing X1 of phase i.

Generalize the steady state balance to apply to this situation.

9. Gas Turbine Efficiency

A high efficiency gas turbine with an inlet gas temperature of 2300 deg F

and an outlet temperature of 1000 deg F is reported to have an efficiency

of 48 %.

Find the maximum efficiency of this turbine, if the ambient temperature

is 60 deg F.

10. Entropy Production in a Cylinder

For a long hollow cylinder conducting heat steadily with uniform surface

temperature T1 on the inner surface of radius r1 and temperature T2 on the

Heat Transfer

surface of radius r2, show that the irreversibility per unit length is given by:

(2k Pi /T1T2) { (T1 – T2) (T1- T2)/ log r2/log r1.

11. Maximum Work Potential of the Mississippi River

In an article in "Science" it has been suggested that the difference in salinity of the Mississippi River and the gulf stream can be used to produce

energy by running a reverse osmosis plant in reverse. If the mole fraction of the salt in the gulf is 0.03 and that in the river is 0.005, estimate the maximum amount of work that can be obtained by this process per kg/sec

of flow. The temperature is 20 degrees C. Assume that the salt is $MgCl_2$ and

that ideal solution theory is true.

12. Thermodynamics

Consider diffusion of water vapor through a membrane 1 mm thick.

The pressure is 0.1 MPa and the temperature is 300 degree K.

The mole fraction of vapor at one end is 0.02 and at the other end it

is 0.01. The diffusion coefficient through the membrane is 0.1 square cm/sec.

(a) Find the local rate of entropy production at the more dilute face of

the membrane. Express the answer in watts/cubic meter-degree K.

(b) Now consider heat conduction through the membrane with a temperature of 10 degrees K. The thermal conductivity of the membrane is very low and is equal to 3.8 x 10 – 3 watts/cubic m-degree K.

(c) Based on the values loss of availability relative to 300 degrees K in parts a and b would you choose the thermal or diffusion methods of separation? The values of physical properties and gradients given are typical for gaseous mixtures.

13. Vapor Compression Cycle Problem

Consider the idealized basic vapor compression cycle. The refrigerant is
Dichlorodifluoromethane. (Refrigerant 12; use tabulated properties).
The surroundings are at 300 degrees K. The cooling load is air which is
cooled from 300 degrees K to 280 degrees K, with a pressure drop of 0.01 MP.

In the idealized system, the condenser operates at a constant pressure
drop of 0.01 MP, and the condenser operates at a constant pressure of 1 MPa
and the evaporator at 0.3 MPa. The pressure drop in the refrigerant is
neglected. An adiabatic expansion valve connects the condenser to the
evaporator. Condensation proceeds to the saturated liquid line and
evaporation to the saturated vapor line.

The refrigerant is recirculated at a rate of 1 kg/sec.
(a) Compute and tabulate the inlet and the exit enthalpies and entropies for
the compressor, condenser, valve and for the evaporator.

(b) Find the entropy production and the availability consumption for the
valve and for the condenser.

Heat Transfer

COP (Coefficient of Performance)

COP of Heat Driven Cycles

The ambient temperature is TA with the inlet flow of Qc Btu/min .The room temperature is TR. A hot fluid ,Th flows at a rate Th.

COP = Qc/W = TR/(TA – TR)

As an example, let TR = 299 K , TA = 308 K and 353 K.

This gives COP = 2.3 .

If Th = 423 K, COP = 4.9.

APPENDIX TWO

In this example we use the properties of saturated refrigerant 12 , which lists the pressure P, MPa, the temperature T, K the volume v, m3/kg , the enthalpy , kJ/kg and the entropy , S, kJ/kg K.

(a) Entropy Production in a Condenser

Condenser

.... Inlet <

....

Evaporator

105   210

Load = m (210 – 105)

(b)   Condenser

Heat in = 250   S out = 0.410 KJ / Kg – deg K

....

....

Heat out = 105

S = 0. 87 KJ/deg K

Entropy Balance

S = m (Sout – Sin) + Q/T

S = m (0.41 – 0.87) – m (250 – 105)/300 =

2.14 times 10 minus 4 KJ/sec-deg K.

(c) Valve

Inflow = one MPa   S = 0.410

Outflow = 0.2 MPa S = 0.426

Entropy Production/m = (Sout – Sin) = 0.4259 – 0.4103 = 0.0156

KJ/Kg-Deg K.

(d) Entropy Production in the Evaporator

....

Heat in = 105.1 KJ/Kg

Heat out = 210.92 KJ/Kg

Heat Transfer

....

Rate of Entropy Production /m = (Soot − Sin) − (hout − hin)/Tcold = (0.8368 − 0.4259) - (210.92 − 105.11)/270 =

0.0190 KJ/Kg − deg K.

m = 0.00001.796 Kw/sec

Rate of Entropy Production = (0.8368 − 0.4259) − (210.92 − 105.11)/270 = 0.0190 KJ/ Kg deg K.

m = 0.00001796 Kw/sec . Entropy production X T0 = 0.00001796 X 300 = 0.005388 Kw.

Fraction of Carnot work = 0.0539/0.208 = 26 percent.

ENTROPY BALANCES

1. What is the maximum amount of work that can be extracted from two bodies initially at temperatures T1 and T2 ?

   Their equations of state are :

   U = NCvT

   where U = internal energy, Cv = molar heat capacity, U = number of

   moles, T = temperature.

   HINT : Make an energy balance in terms of the final temperature, Tf.

   Find Tf from dS = dQ1/T1 + dQ2/T2 = 0.

   This is the reversible energy balance on the engine.

Heat Transfer

2. Two identical bodies of constant total heat capacity of 10 kcal/0K each are at the same initial temperature To. A refrigerator operates between these two bodies until one body is cooled to a temperature T1.

    If the bodies undergo no change in phase and remain at a constant pressure, find the minimum amount of work required for this process in terms of T0 and T1 .

3. A home is to be maintained at 70 degrees F , and the external temperature is 50 degrees F.One method of heating the home is to purchase work from a power company and to convert it directly to heat. This method is used in electric room heaters. Alternatively, the purchased work can be used to operate a heat pump.

4. What is the ratio of costs if the heat pump attains the ideal thermodynamic coefficient of performance?

    Reference H.B. Callen," Thermodynamics and an Introduction to Thermostatics" John Wiley and Sons, NY, 1985.

5. Entropy Production due to Mixing.

Component A flowing at a rate of ma moles/sec is mixed with component B which at a rate of mb moles/sec. Find :

(a)   the entropy production due to mixing neglecting pressure drop and any heat evolution. Assume the mixture obeys the ideal solution theory. The process has reached a steady state.
(b)   the entropy production with a pressure drop of delta P Pascals, assuming the mixture is incompressible and has a density of rh Kg/cubic meter and a molecular weight M.
(c)   the entropy production with a heat leak of q J/sec to surroundings at a temperature of To degrees K.

Heat Transfer

(d)   What is the relation of the entropy production in part "a" to the
        minimum work needed to separate the mixture into pure components A
        and B ?

(5) Availability Loss in Cooling Water

Cooling water leaves a condenser of a power plant at a temperature Tc and is
thrown into a large lake at a temperature To. Fresh cooling water is taken from
the lake at the same flow rate ma kg/sec.The specific heat of the water is a
constant equal to Cp J/Kg-deg K.Find the loss of availability of the hot water
dumped into the lake in terms of ma,Cp, To,and Tc.

(6) Thermodynamic Efficiency of Distillation

A distillation column separates a 50-50 mixture of propane and propylene into
two products, each 99% pure. The column is heated with steam at 212 degrees F
and cooled with water at 70 degrees F. The entropies of feed, the distillate, and
the bottoms have been calculated using the equation for the mixture :

S = - R (Sum of x1lnx1)

where S1 is expressed in terms of Btu/lb-mole-degree R.

(a)   Using the heat input values calculate the entropy changes for the steam
        and the cooling water , assuming a negligible rise temperature .
(b)   Make an entropy balance and calculate the thermodynamic efficiency.

(c) How can you make this column more efficient ?

(7)   Second Law Analysis of Counter- Current Membrane Mass Exchangers

        Consider hydrogen separation through a membrane and transfer to a
        carbon dioxide stream. Assume that the hydrogen obeys the

        ideal solution theory and that the temperature is To .

17

(a)   Find the minimum work necessary to separate hydrogen from the inlet mole fraction of xH2.

(b)   Find the entropy production and the availability consumption due to a pressure drop of delta P pascals along the length L

of the hydrogen – nitrogen stream .

(c) Find the entropy production due to a pressure drop delta P across the membrane of thickness "l".

(d) Formulate a linear flux-force for membrane flow and diffusion in terms of delta P and delta x H2.

(e) Find the entropy production for the exchanger and explain why counter-current operation is desirable.

(8)   Thermodynamics

Friction, F can be defined by means of the steady state mechanical energy balance ,

$vdv + gdz + dP/rho = - dF - dW$

where F acts as the energy that has been added to the Bernoulli equation to make it exact.

For pipe flow F is expressed in terms of a friction factor, f defined by means of the equation

$F = 2fvvz/D$.

(a)   Use the steady state energy balance
       $d(u + P\ rho-1 + vv/2 + gz) = dQ – dW$ to find an

expression for TdS in terms of the friction factor, f.

(b)   Consider flow of a multiphase mixture containing x1 of phase i. Generalize the steady state energy balance.

Heat Transfer

NOTATION

D pipe diameter

f friction factor

F friction

g gravitational constant

P pressure

Q heat into the system

S entropy

T temperature

v velocity

W shaft work

x I weight fraction of i

rho density

z height

u   internal energy

9. Gas Turbine Efficiency

A high efficiency gas turbine with an inlet gas temperature of 2300 degrees F

and an outlet temperature of 1000 degrees F is reported to have an efficiency of 48 per cent .(Reference : Bajura and Webb, Mechanical Engineering ,p 58 Sept 1991)

Find the maximum efficiency of this turbine, if the ambient temperature for cooling is 60 degrees F.

10. Entropy Production in a Cylinder

For a long hollow cylinder conducting heat steadily with uniform surface temperature T1 on the inner surface of radius r1 and temperature T1 on the inner surface of radius r1 and T2 on the outer surface of radius r2 , show that the irreversibility per unit length is given by :

2pik/T1T2   X (T1 – T2)(T1 – T2)/lnr2/r1 .

Hint for the cylinder:

Q= 2 pi rq, where q= - k dT/dr

T = (T2lnr/r1 – T1lnr/r2)/ lnr2/r1

11. Maximum Work Potential of the Mississippi River

In an article in Science by professor Levenspiel it has been suggested that the the the salinity of the Mississippi River and the gulf stream can be used to produce energy by running the reverse osmosis plant in reverse. If the mole fraction of the salt in the gulf is 0.03 and that in the river is 0.005, estimate the maximum amount of work that can be obtained by this process per kg/sec of flow.The temperature is is 20 degrees C. Assume that the salt is $MgCl_2$ and that the ideal solution theory is true.

12. Diffusion of Water Vapor Through a Membrane

Consider diffusion of water vapor through a membrane 1 mm thick. The pressure is 0.1 MPa   and the temperature is 300 degrees K. The mole fraction of vapor at one end is 0.02 and. at the other end it is 0.01. The diffusion coefficient through the membrane is 0.1 cm square/sec.

(a)   Find the local rate of entropy production at the more dilute face of the membrane. Express the answer in watts/cubic centimeter degree Kelvin.

Heat Transfer

(b)   Now consider heat conduction through the membrane with a temperature drop of 10 degrees Kelvin. The thermal conductivity of the membrane is very low and is equal to 3.8 x 10 -3 watts/m-degree Kelvin. If one face of the membrane is 300 degrees Kelvin, find the local rate of entropy production at this face in terms of watts/cubic meter -degrees Kelvin.

Based on the values of loss of availability relative to the 300 degrees Kelvin in the other parts,   would you choose the thermal or the membrane diffusion methods of separation ?

The values of physical properties and gradients given are typical for gaseous mixtures.

VAPOR COMPRESSION CYCLE PROBLEM

Consider an idealized basic vapor compression cycle. The refrigerant is discholorodiflouromethane (refrigerant 12. Use tabulated properties.

The surroundings are at 300 degrees Kelvin. The cooling load is air which is cooled from 300 degrees Kelvin to 280 degrees Kelvin, with a pressure drop of 0.01 MP.

In the idealized system, the condenser operates at a constant pressure of 1MPa and the evaporator at 0.3 MPa. The pressure drop in the refrigerant is neglected. An adiabatic expansion valve connects the condenser to the evaporator. Condensation proceeds to the saturated liquid line and evaporation to the saturated vapor line. The refrigerant is recirculated at a rate of 1 kg/sec.

Heat Transfer

(a)   Compute and tabulate the inlet and the exit enthalpies and entropies for the compressor, condenser, valve and for the evaporator.

(b)   Find the entropy production and the availability consumption for the valve and for the condenser.

Heat Transfer

A Short Biographical Sketch
Dimitri Gidaspow

Education:

B.Ch.E. (cum laude), the City College of New York, 1956

M.ChE. Polytechnic Institute of Brooklyn, 1959

Ph.D., Illinois Institute of Technology, 1962

Academic Experience:

Distinguished Professor of Chemical Engineering, IIT; University Professor, 2000-Present; Professor, 1977-00; Adjunct Assistant Professor to Full Professor, Institute of Gas Technology (IGT), 63-77; Principal Advisor of 50 Ph.D. students. Major graduate courses taught in the last 10 years: transport phenomena, fluidization, heat transfer, computational techniques, thermodynamics.

Industrial Experience:

Air Product 1956

IGT - helped to develop two new energy technologies: (1) fuel cells; summarized in a dozen refereed papers, in 2 patents and recognized by NASA by means of a 1969 award from the Marshall Space Flight Center, and (2) desiccant air conditioning; documented in 12 papers and a patent issued in 1982.

AEC National Reactor Testing Station in Idaho - consultant 1972-1974. Consultant to Lawrence Livermore Laboratory, Energy Research Corp., Argonne National Lab.; Department of Energy (fuel cell survey), UTC, EXXON, Westinghouse and IITRI, MOBIL, Dallas and UOP.

Major Professional Activities:

Papers Chairman for AIChE Heat Transfer and Energy Conversion Division, 1973-1974. Editor, "Heat Transfer - Research and Design", AIChE Heat Transfer Symposium Series Volume 70, No. 138 (1974). Program Chairman for the 1979 Intersociety Energy Conversion Engineering Conference and editor of a 2 volume set of proceedings (ACS). Chairman (1981-84), AIChE multiphase flow committee (7g). Particle technology chairs at AIChE annual meetings, last 5 years.

Publications and Patents:

9 patents & 2 pending. 185 publications in AIChE Journal, Chemical Engineering Science, I&EC Fundamentals, Symposium Series, etc. Multiphase Flow and Fluidization, Continuum and Kinetic Theory Description book, published by Academic Press (1994).

Heat Transfer

Awards:

AIChE 1984 Donald Q. Kern Award presented at the 1985 Heat Transfer Conference. Lecture: "Hydrodynamics of Fluidization and Heat Transfer: Supercomputer Modeling", in Appl. Mech. Rev. 39, No. 1, 1-23, 1986. NSF 1986 Creativity Award for work in separation of particles in non-aqueous media with D. T. Wasan. Fellow of the American Institute of Chemical Engineers. IIT Alumni award. AIChE 2002 Flour-Daniel Lectureship Award in Publications Relevant to Present Research

1. Gidaspow, D., "Multiphase Flow and Fluidization: Continuum and Kinetic Theory Description", Academic Press, 1994

2. "Hydrodynamics of fluidization using kinetic theory: an emerging paradigm. 2002 Flour-Daniel lecture", Dimitri Gidaspow, Jonghwun Jung, Raj K. Singh, Powder Technology, 148, 123-141 (2004)

3. " Kinetic theory based CFD simulation of turbulent fluidization of FCC particles in a riser", Veeraya Jiradilok, Dimitri Gidaspow, Somsak Damronglerd, William J. Koves and Reza Mostofi, Chemical Engineering Science 61, 5544-5559 (2006)

4. "A Bubbling Fluidization Model Using Kinetic Theory of Granular Flow", J. Ding and D. Gidaspow, " AIChE Journal 36, 523-538 (1990) (283 citations)

5. "Computations of Flow Patterns in Circulating Fluidized Beds", Y.P. Tsuo and D. Gidaspow, Presented at the 26th National Heat Transfer Conference, Philadelphia, August 1989: AIChE Journal 36, 885-896, 1990

Five Other Significant Research Publications

1. " Measurement of Granular Temperature and Stresses in Risers", Mehmet Tartan and Dimitri Gidaspow, AIChE Journal 50, 1760-1775 (2004)

2. "Multiparticle simulation of collapsing volcanic columns and pyroclastic flow",

A. Neri, T.E. Ongaro, G. Macedonio and D.Gidaspow, Journal of Geophysical Research,

Vol. 108, No.B4, doi:10.1029/2001JB000508 (2003)

3. "A Model for Discharge of Storage Batteries", D. Gidaspow and B. S. Baker, Extended Abstracts for the Miami Beach Electro-chemical Society Meeting, Oct. 8-13 (1972), pp. 49-50 J. Electrochem. Soc. 120, 1005-1010 (1973)

4. 1984 Donald Q. Kern Award presented at the 1985 Heat Transfer Conference. Lecture: "Hydrodynamics of Fluidization and Heat Transfer: Supercomputer Modeling", in <u>Appl. Mech. Rev.</u> 39, No. 1, 1-23, 1986

5. "Nonlinear Coupled Heat and Mass Exchange in a Cross-Flow Regenerator", R. Dipak and D. Gidaspow, <u>Chem. Eng. Science</u> 29, 2101-2114 (1974)